SEASHORE
LIFE

BY

BARBARA TAYLOR

WHERE LAND MEETS THE SEA

F rom spectacular, wave-battered rocky shores and hot sandy beaches fringed with coral, to muddy estuaries and mangrove forests, coastlines are fascinating meeting places between two very different environments. Most shoreline plants and animals are sea inhabitants that have adapted to living out of water for part of every day. But a few, such as insects high on the shore, plants on sand dunes, and birds that feed on the shore or nest on cliffs, are land dwellers. The twice-daily ebb and flow of the tide exposes coastline creatures to dramatic changes in temperature and water levels. These changes are as great as seasonal changes on land, but they take place every day instead of every year, making this a very challenging habitat for wildlife.

HERMIT AT HOME

Hermit crabs look ungainly as they crawl over the shore scavenging for dead animal remains. Yet they are well suited for shoreline survival. The large front pincer can be wedged across the opening of the shell, sealing the entrance like an armor-plated door. Then the crab can be rolled about by the waves without damage to its soft body inside the shell. The crab's body also remains moist inside the shell when the tide goes out and it is left exposed to the air.

BEACH BURROWERS

A sandy beach often seems to be empty of life, but most of its inhabitants are buried beneath the surface. These range from microscopic bacteria to all sorts of shellfish, crustaceans (such as crabs), worms, and sea urchins. They trap food particles from the water brought in by the tides or extract nourishment from the sand, rather like an earthworm eats soil.

FOSSIL FINDS

Fossils found in rocks on beaches may help to date the rocks or provide evidence of changing sea levels. Fossils of ocean creatures, such as this ammonite, show that the rocks were once under the sea. Ammonites were shellfish related to modern-day octopuses, squid, and nautiluses. When the animal was alive, a head with tentacles stuck out of the end of the shell, but it could retreat back inside and close the opening. Ammonites lived hundreds of millions of years ago.

HIDING PLACES

A rocky shore has a variety of places for animals to feed and shelter. They can hide under rocks and boulders, among tangled masses of seaweed, or in rock pools. Some creatures, such as piddocks and sea urchins, even bore themselves into the solid rock. Rocks extending out from the shore can be hazardous for shipping, but a lighthouse warns ships to stay well clear at night.

CONE PROTECTOR

Limpets are supreme masters of clinging to rocks with their muscular foot. They even hollow out a shallow pit in the rock for a better grip. This is their home base. The cone shape of the limpet's shell gives it a wide base for clinging to rocks with minimum resistance to the water. Waves and tides wash over the limpets without dislodging them. Limpets usually move about when the tide is in, grazing on algae on the rocks. They return to their home base when the tide goes out.

CLIFF COLONIES

Many seabirds nest on rocky cliffs where it is difficult for predators to reach their eggs and where they are close to their food supply in the sea. Puffins live in complicated nest burrows on cliff tops and have to choose sites with enough soil. The large, colorful bill of the Atlantic puffin plays an important role in pair formation and courtship. Outside the breeding season, the puffin's bill is smaller and less brightly colored.

LIFE BETWEEN THE TIDES

The lives and activities of shore creatures are controlled largely by the rhythm of the tides, which creep up the shore and then down again, roughly twice every 24 hours. In some parts of the world, such as the Mediterranean, the tides are very small, but in other places the tides go out as much as 2 miles (3 km). Creatures living between the tides have to survive in two environments, both in and out of the water. Creatures that can survive out of the water for long periods live high up the shore. Those that can only tolerate short periods out of the water live on the lower shore, near the sea. There are often distinct zones or bands of life, from the sea to the top of the shore. At low tide, shore creatures are exposed to drying winds, the sun's ultraviolet radiation, high and low temperatures, the fresh water in rainfall, and attacks from land-based predators.

SURFING SNAIL

The South African plow snail has perfected a remarkable way of moving with the tides. As the tide moves up the shore, the plow snail emerges from its hiding place in the sand and sucks water into its foot. This makes a plow-shaped "surf board" on which the snail surfs up the beach. At the high-water mark, where the waves deposit all sorts of dead and decaying food, the snail feeds on stranded sea creatures. As the tide retreats, the plow snail surfs back down the beach.

Barnacles out of water

HEADSTANDS

Barnacles float in the water for a while before settling down on the rocks. They are attracted by the scent of existing colonies and tend to settle near them. It is said that a barnacle "welds its head to the rock and spends its life kicking food into its mouth with its legs." When the tide is out, the barnacle pulls its legs back inside its shell and seals the opening with six stout, limy plates.

A barnacle under water

WHY TIDES HAPPEN

THE MOON'S ORBIT

THE EARTH

The tides are caused mainly by the pull of the moon's gravity on the Earth's oceans. As the moon orbits the Earth, the oceans are pulled towards it, making high tides on that side of the Earth – and on the opposite side because of the Earth's spin. The sun's gravity also pulls the oceans, but its effect is much weaker because it is much farther away. Twice a month, when the sun, moon, and Earth are in a line, the pull of the moon and sun combine to produce extra high and low tides, called spring tides. When the sun, moon, and Earth form a right angle, the sun's pull works against the moon's pull, making less extreme tides, called neap tides.

BURROWING SHELL

Cockles live on sandy shores, burrowing into the sand with their muscular foot. They only burrow deep enough to cover the shell and may even "leap" and roll over the sand to change their feeding area. The ribbed shell probably helps to hold the animal within the sand. When the tide is in, cockles push two short siphons above the sand to filter tiny food particles, such as plankton, from the water. Cockles survive best in mid to low shore levels where the tide flows rapidly to and fro. Over 10,000 cockles can live in just 10 sq ft (1 sq meter).

BEACH STAR

Starfish usually live below the level of the lowest tide or in rock pools. If they are washed up on a beach, they will dry out and die. To avoid this from happening, starfish have hundreds of tube feet to cling tightly to rocks, sand, and other surfaces. They are also protected by an external skeleton of tough, limy plates embedded in the tough skin. If a starfish's arm is crushed by a waveswept boulder, or bitten off by a predator, it can grow a new arm.

COASTLINES OF THE WORLD

Coastlines are very special and varied places. They range from the sun-baked tropical shores of Australia and Indonesia to the temperate shorelines of Europe and the frozen coasts of Arctic Canada. Some, such as the shores of the Galapagos Islands, are a unique home to wildlife that is found nowhere else in the world. Coastlines vary tremendously with the climate in different parts of the world. Coral reefs, for instance, grow only in warm, clean waters in the tropics. Mangrove trees are also characteristic of tropical shores and are adapted to grow in seashore mud in the tidal zone. In cooler parts of the world, such as Northern Europe and North America, sandy, rocky, and muddy shores are less exotic, but still full of creatures with extraordinary adaptations. The muddy shores of European estuaries are particularly important as feeding and resting places for migrating birds. Polar shores come alive in summer, as feeding and nesting areas for whales, seals, and millions of seabirds.

EUROPEAN SHORE

Life on northern European shores is very exposed to the sun, the wind, and the waves. Yet rocks and seaweeds provide shelter for a surprising variety of creatures and a little way beneath the sand a host of burrowers and tube dwellers survive in a relatively constant environment. However, the weather does change with the seasons and winter storms may pound these coasts with considerable force, removing old or poorly attached creatures.

CORAL COASTS

For hundreds of millions of years, coral reefs have grown along tropical shores where the sea temperatures are always above 64˚F (18˚C). They grow only in shallow water up to 230 ft (70 meters) or so deep because the algae that live in the coral need sunlight to make their food. The reef itself is made of the skeletons of coral animals (which are rather like sea anemones), with living coral on top. In ideal conditions, healthy reefs grow up to 1 inch (25 mm) a year.

ORCA COAST

These striking dorsal fins belong to killer whales, or orcas, living off the coast of British Columbia, Canada. These toothed whales will chase fish and other prey into quite shallow coastal waters. Killer whales live in oceans all over the world, but are most common in cooler climates. They are a top predator around polar coasts. Whales of all kinds breed near tropical coasts but migrate to polar regions to feed there in summer.

MANGROVE COASTS

Mangrove, or tidal forest, grows on river estuaries and sheltered muddy inlets in the tropics. Forming a living barrier between the land and the sea, the intricate network of mangrove roots protects soft-soiled coasts from flooding or being washed away by the waves and the tides. A wide variety of animals, such as fish and crabs, live in the shelter of the trees.

GALAPAGOS COASTS

From sea lions and marine iguanas to flightless cormorants and frigate birds, the coasts of the Galapagos Islands are a rich treasure trove of unique wildlife. Sea lions breed on most of the islands, which are isolated in the Pacific Ocean about 620 miles (1,000 km) west of South America. Warm waters from the Pacific and cool waters from the Antarctic both flow past the Galapagos. This means that animals from cold places, such as penguins, live side by side with tropical species, such as flamingos.

ROCKY SHORES

T he most obvious plants on rocky shores are the red, green, and brown seaweeds. They have thick outer layers and a low surface area to cut down on water loss when they are exposed to the air. Most successful animals of the high shore have shells, which may be white and tall to reflect heat in warmer climates. Shellfish resist water loss at low tide by keeping their shells tightly closed or by clamping them firmly to the rocks. They may also be able to breathe air as well as take in oxygen from the water. Animals that cannot stand being exposed to the air, such as sea anemones, crabs, worms, and sponges, have to shelter among rocks or in tide pools as the tide goes out.

HIGH TIDE ZONE

MIDDLE TIDE ZONE

LOW TIDE ZONE

ZONES OF LIFE

Plants and animals live in zones on all intertidal beaches, but the zones are most obvious on rocky shores. There are three major zones: the low tide zone, with animals such as starfish and sea squirts and seaweeds such as kelps; the middle tide zone, with animals such as barnacles and mussels, and seaweeds such as wracks; and the high tide zone, with animals such as periwinkles and limpets, as well as lichens. Plants and animals are restricted to a zone according to how long they can survive out of the water.

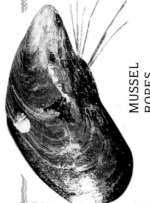

MUSSEL ROPES

Mussels anchor themselves to rocks or even other mussels, using strong, sticky byssal threads, which work like the guy ropes on a tent. The mussels produce the threads as a thick fluid that hardens in the seawater. Mussels depend on the waves and tides to bring their food. They open their shells slightly under the water, and filter microscopic plants and other food in.

SEALED SHELL

On the back of this common whelk's foot is a horny operculum that seals the opening of the shell when the whelk retreats inside. The shell is also thick and ridged to help the whelk survive the pounding of the waves. Whelks lay their eggs in a spongy ball, which may be washed up high on the beach along the strandline.

SANDY SHORES

S andy beaches support fewer species than rocky shores, although some animals, such as worms and bivalve mollusks, may exist in immense numbers. Buried beneath the sand is a thick carpet of wriggling bodies. As many as 8,000 burrowing clams have been counted in only 10 sq ft (1 sq meter). A few inches below the sand, conditions are much the same whether the tide is in or out, or whether it is warm or cold, sunny or raining. A film of water surrounds each sand grain, sticking them together so that the sand is always moist, even up to the high tide mark. Burrowing creatures feed on organic debris brought in by the tide, but there are also predators, such as the masked crab, the burrowing starfish, and seashore birds. At the high water mark, where bits of seaweed, shells, and other debris collect on the strandline, scavengers such as sandhoppers and turnstones find plenty to eat.

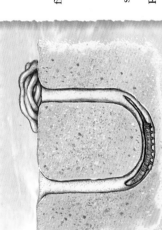

SAND CRABS

At low tide, the sandbubbler crab sifts small pieces of food from the sand, leaving the cleaned sand behind in little round balls. Other sand crabs are usually buried just beneath the surface, ready to filter food particles from the water. Ranging in length from 1–3 inches (2.5–8 centimeters), they move up and down the beach near the shoreline.

CUNNING GULLS

Gulls sometimes paddle their feet up and down on the surface of the sand to bring cockles and other hidden animals to the surface. They may also carry shellfish away to a hard surface and drop them so they smash open. Gulls are real scavengers and their strong beaks can usually deal with most kinds of food. They can even crack open the shells of crabs to get at the juicy meat inside.

BURROWING ANIMALS

At low tide, there is often evidence of life beneath the surface. Squiggly mounds of sand, rather like piles of spaghetti, are the waste sand squirted out the end of a lugworm's burrow when it has finished feeding. The grainy feeding tubes of the sand mason worm may also be sticking out of the sand. The worm makes this protective tube from sticky mucus to which grains of sand become attached.

SAND DUNE SECRETS

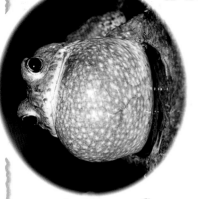

S and dunes sometimes build up along a shore where windblown sand is anchored by tough plants such as marram grass. The unstable dry sand of the dunes contains little organic material and the surface may become baking hot on a sunny day. Yet some creatures still live here. Wolf spiders survive by hunting flies blown into the dunes from the strandline. The spiders are eaten by the rare and beautiful sand lizard. Rabbits find it easy to burrow in the dunes and there are plenty of plants for them to eat. Shelducks may use old rabbit burrows as nesting sites. Other birds, such as skylarks, terns, gulls, and plovers also nest on the dunes, although foxes may raid their nests. Sheltered damp hollows between the larger dunes, known as "slacks," contain orchids, rushes, and other marsh plants. The dune slacks provide breeding grounds for frogs and toads.

TERRIFYING TOAD

The rare natterjack toad lives in dune slacks. It puffs out its vocal sac to amplify its calls, which usually begin just before sunset. On still, quiet evenings, a chorus of several toads can be heard over 1¼ miles (2 km) or more. When it is disturbed or alarmed, this toad inflates its body, raises its rump and produces a horrible smelling secretion from its skin.

LAZY LIZARD

Sand lizards are cold-blooded so their bodies are always at the same temperature as their surroundings. They need the warmth of the sun to give them energy and keep their bodies working. Sand lizards lay their eggs in the sand, using the warmth of the sand to help the development of their eggs.

MIGHTY MARRAM

Marram grass has tough leaves rolled into tubes that trap valuable moisture. It can put out roots from its stem, enabling it to climb "step-by-step" up the growing dunes, binding the sand together as it goes. Its roots penetrate deeply into the sand to reach fresh water in the cold sand deep below. When a shoot is buried by the sand, it produces a bud that pushes up another vertical shoot. In one year, marram grass can spread up to 29 ft (9 meters) horizontally and 3 ft (1 meter) vertically, helping the dunes to grow in the process.

DUNE PLANTS

Dune plants have several features that help them survive the gale-force winds, the drying heat, the lack of moisture, and the wind-blown salt. Sea holly (right) and sea bindweed have stiff "varnished" leaves to stop water from escaping and give protection against sharp sand grains blown by the wind. Hairy plants trap dew on their stalks and leaves. Many plants grow in rosettes that hug the ground to keep out of the wind. Others have deep and extensive root systems to anchor them securely and reach water underground.

POISONOUS MOTHS

Day-flying cinnabar moths feed from dune flowers and their striped caterpillars (right) are often seen feeding on plants such as ragwort or groundsel. The red warning color of the moth is the same color as the pigment known as vermilion or cinnabar. The striking colors of both the moth and the caterpillar warn of their poisonous bodies and protect them from predators such as lizards, birds, and mammals. The poisons come from the caterpillars' food plants and are passed on to the adult moth.

LIFE IN A ROCK POOL

Rock pools are like natural aquariums on the shore. Shallow pools only have a few plants and animals, but deeper ones may be packed with life. Rock pools provide permanently wet refuges at all levels on the shore. Yet the environment of a rock pool is tough. Its inhabitants have to withstand great swings in temperature, salinity, and acidity but changes in the level of oxygen and carbon dioxide are often most dramatic. At night, when the plants are not photosynthesizing (or making food), carbon dioxide builds up but oxygen levels fall as both plants and animals continue respiring and using up oxygen. The extra carbon dioxide created at night makes the water acidic. The reverse happens during the day as plants use up carbon dioxide and give off oxygen during photosynthesis. There is a constant back and forth of life between the rock pools as animals search for food or seek to avoid predators.

SEAWEED SURVIVAL

The brown seaweed called bladder wrack has flat, wave-resistant fronds and air-filled bladders on its fronds. The bladders buoy up the seaweed when the tide is in so they float off the rocks, with their branches well apart for photosynthesis. A slimy covering protects the seaweed from drying out at low tide, when air and water are trapped between the layers of fronds. The seaweed is firmly attached to the rocks by a long, flexible stalk. The orange fruiting branches contain the spores that will grow into new seaweeds.

JELLY FLOWERS

Sea anemones flourish in rock pools because they are usually underwater and can feed all day. They may look like flowers but they are hollow, jelly-like animals related to jellyfish and corals. Their colorful waving tentacles are a forest of danger for small sea creatures, such as shrimps, which are stung by the tentacles and pulled in towards the mouth. To protect themselves from danger, or drying out, most anemones can pull in their tentacles and become jelly-like blobs.

12

POOL PRAWNS

Most rock pools have a thriving shrimp and prawn population but, being almost transparent, they are not easy to see. The chameleon prawn can change color to match its surroundings, although it takes about a week to do so. Prawns can also shoot backwards suddenly to avoid danger. They can live in a wide range of temperatures, but move to deeper, warmer water at the beginning of winter.

STAR TURN

The common starfish has a very effective way of eating mussels. First it wraps its arms around the shell and pulls hard until the exhausted mussel allows the two halves of its shell to open a little. Then the starfish turns its stomach inside out and slips it down inside the mussel to slowly dissolve and devour the contents.

ROCK POOL FISH

Rock pool fish, such as blennies and gobies, are usually well camouflaged and shaped so that they can easily wriggle in among rocks and seaweed. Their eyes are often near the tops of their heads to watch for predators from above. Lumpsuckers and clingfish have suckers to hang on to rocks when waves crash over them or the tide drains water from the pools. In spring, shore fish come to rock pools to breed. The young move to warmer offshore waters for the winter.

LIFE IN A ROCK POOL

This rock pool in South Africa has an abundance of life – creatures and plants all crammed together in their wet refuge.

The empty shell, or test, of a sea urchin. The white knobs show where the spines were attached.

Starfish crawl into rock pools to find damp places to shelter in low tide.

The conical shape of the limpet's shell helps it resist the pounding of the waves.

Crabs hide in rock pools to stop their bodies from drying out. They sort through the debris for food particles.

At high tide, sea urchins feed on tiny plants by scraping rocks with their powerful teeth.

Sea lettuce is green seaweed that looks like the lettuce we eat in salads.

GANNET LIFTS

Gannets and other seabirds soar on currents of air that rise up over the cliffs. Gannets are strong fliers, but this saves energy at a time when they have to make many trips to and from the nesting sites to gather food for their young. Gannets use their strong pointed bills to catch fish and also to stab at other gannets that come close to their nesting space. They build large nests of seaweed, feathers, grass, and soil.

PLASTER NEST

Kittiwakes plaster their cup-shaped nests to the sheer rock face with mud and droppings. They can nest on ledges that are too narrow even for guillemots. It is usual to see only two young in the nest until they fly at about six weeks old. Kittiwakes are named after the sound of their call.

EGG ROLL

Guillemots do not make nests. Their eggs are pear shaped and if they are blown by the wind or touched by a bird they spin around rather than fall off the cliff. Each guillemot has a different pattern on its egg to help it to recognize its own. Some have rust-red stripes, others have chocolate blotches and some have green and black squiggly "writing" all over them.

LIFE ON A LEDGE

The sight of hundreds of seabirds swooping and screeching around a cliff nesting site is unforgettable. The birds nest at different levels to divide up the living space (below right). Near the top of the cliffs nest gannets, gulls, and puffins, with guillemots, razorbills, and kittiwakes in the middle, and larger cormorants and shags at the bottom. The steepness of the rocks keep the eggs and chicks safe from most predators. Also the young birds are close to the sea when they are ready to leave the nest. Cliff plants cannot follow the birds to warmer climates when the summer breeding season is over. Instead, they must complete their life cycles in constant salty spray, come rain, frost, snow, or landslide. Long, probing roots help them to hold on to thin cracks in the rock, and their fleshy leaves and waxy surfaces reduce water loss.

GROUND FLOOR

Shags and cormorants nest at the bottom of cliffs, building untidy nests of sticks and seaweed, lined with grass. Shags defend their nests vigorously. They refuse to leave during an attack and thrust their beaks forward at intruders. After swimming, shags and cormorants often stand with their wings open to let their feathers dry in the breeze.

VEGGY CLIFFS

Ancestors of well-known vegetables grow on cliffs, such as sea carrot (left), sea cabbage, and sea beet. Sea beet is quick to colonize newly broken ground so it grows well in the unstable cliff environment, where pieces of cliff may break off and slide down into the sea. In the breeding season, birds kill the plants by trampling on them and covering them with their droppings. But some droppings provide nutrients for plants such as sea campion. Tree mallow grows more luxuriantly among a jumble of herring gull and cormorant nests.

GULLS

PUFFINS

GANNETS

KITTIWAKES

GUILLEMOTS

RAZORBILLS

CORMORANTS

SHAGS

WHERE RIVERS MEET THE SEA

When the freshwater of rivers slows down and flows into the salty water of the sea, the mud and silt it carries clump together and drift to the bottom. Over hundreds of years, the bare mud builds up, plants grow, and a saltmarsh forms. The main problem with living in these muddy estuaries is the huge variation in salinity caused by the tides, and variations in river flow due to rain or drought. If a sea creature, with blood as salty as seawater, is placed in freshwater, it will absorb water, swell up and die. Twice a day, most of the mud will be exposed to the air and there are also dramatic variations between summer and winter. Estuary inhabitants must be able to withstand a great range of chemical and physical conditions but a few creatures, such as worms and shellfish, flourish in huge numbers. A quarter of a million sludge worms can live in just 10 sq ft (1 sq meter). Estuary mud is a rich source of food, delivered daily from both sea and land.

SHARING FOOD

A variety of wading birds can feed together because their bills are different lengths and shapes, enabling them to feed at different depths in the mud. Short-billed plovers feed on spire shells at the surface. Knot and redshank, with bills twice as long, probe into the top layer of mud for shrimps and small worms. Curlews and godwits, with the longest bills of all, probe down deep enough to reach the lugworms and drag them out of their burrows.

WATER CONSERVATION

A pioneering plant called glasswort begins the process of turning bare mudflats into a saltmarsh. It looks rather like a succulent desert plant because its swollen stems also store water and its leathery leaves and stems prevent water escaping. Glasswort needs to conserve water because it cannot take in fresh water from the sea and the very salty seawater tends to draw the water out of its cells. Long ago, people collected glasswort and burned it, because its ashes could be used to make glass.

MUD TO MARSH

Glasswort traps mud, raising the level of the sea bottom and allowing grasses, rushes, and other plants, such as sea aster (left) and sea lavender, to grow. Plant debris becomes tangled among the stems, trapping more mud. The level of the saltmarsh gradually rises until only the highest tides flood it. Only a few ducks and geese graze on the tough and salty plants. Wading birds and ducks nest among the vegetation in summer.

PINCER POWER

Ragworms are active predators, seizing prey in their pincer-like jaws. They live in slime-lined burrows in the mud, reaching out to scavenge on plant and animal remains and catch creatures such as shrimps as they pass by. Ragworms are omnivores and feed on almost anything edible they can find. On each body segment are a pair of flat, paddle-like parapodia. The ragworm uses these for crawling, swimming, and breathing. Stiff bristles on the parapodia help the ragworm to grip surfaces.

FLAT FLOUNDERS

The flounder is a typical estuary fish. It spawns in the sea, but the adults can survive in an estuary or even in freshwater for a long time. Rivers form an important nursery ground for the flounder. They live on the bottom and feed on invertebrates, such as marine worms, small crustaceans, and shellfish. Since they lie on their sides, flounders have both eyes on the right side of their heads.

MANGROVE SWAMPS

SEMAPHORE SIGNALS

One of the male fiddler crab's pincers is much larger than the other and colored pink, blue, purple, or white. He waves this outsize claw at the female, like a semaphore flag, to signal that he is ready to mate. Sometimes, he jumps up and down at the same time. If a female is ready to mate, she trots towards the male of her choice and follows him into his burrow.

A round 93,000 sq miles (240,000 sq km) of sheltered shores in the tropics are fringed with mangrove swamps. Mangroves are trees specially adapted to live in wet, salty, muddy places. They have arching or spreading roots to anchor them in the slushy, glutinous, smelly mud. Some of them "breathe" through small, finger-like roots that stick up through the mud. Breathing roots are vital because there is very little oxygen in the wet mud. The tides regularly cover the base of the trees and a wide variety of animals lives in the mud, on the mangrove roots, and among the branches. The warm shallow waters are an ideal nursery for young fish, which are hunted by crocodiles, alligators, snakes, and birds, such as storks, ibises, and herons. Crabs live in the mud and scuttle about in the jungle of roots at low tide, scavenging for food that is carried in by the tide.

MARVELOUS MANGROVES

Red mangroves support themselves with a tangle of stilt or prop roots. Mud slowly collects around the roots and builds up into soil. So the mangrove trees help to extend the coasts farther out to sea. Their special, torpedo-shaped seeds germinate while still hanging on the tree. When they fall off, some stick like spears into the mud and sprout into new trees. Others are carried out to sea by the tides and may drift hundreds of miles before colonizing another salty patch of coastal mud.

FISH OUT OF WATER

Mudskippers are finger-sized fish that may spend more time out of water than in it. They move across the mud in a series of "skips," by wriggling their tails. They can even climb onto mangrove roots using their stumpy fins. When the tide is out, or when danger threatens, they burrow into the mud. Some mudskippers claim territories, building low mud ridges along the boundaries to keep neighbors out. The mud walls also stop the seawater from draining away at low tide.

CLIMBING SNAILS

Sea snails feed on algae on the surface of the mud at low tide but climb up the mangrove roots before the tide washes seawater over them again. This helps them to escape from fish that might eat them. Every month, the tides rise so high that the snails would not have enough time to get out of harm's way. So they climb even higher, rather than descending to feed (right).

WATER PISTOL

One of the fish that cruises around the mangrove roots at high tide is the archerfish. It spits a stream of water up into the air – like a jet from a water pistol – to knock insects into the water where it immediately gobbles them up.

LOUDSPEAKER NOSE

Male proboscis monkeys have a large, trunk-like nose which probably acts as a loudspeaker for the male's honking calls. These warn other proboscis monkeys of danger. Proboscis monkeys are very agile, leaping through the mangrove forests using their long tails as a counterbalance. If danger threatens, they will dive and swim under the water. Proboscis monkeys live only in the mangrove forests on the island of Borneo. They spend most of their lives high up in the mangroves, feeding on young shoots, leaves, and some fruits and flowers.

HOW REEFS FORM

Coral reefs sometimes grow around the shores of tropical islands. If sea levels rise, or the island sinks, the coral continues to grow upward, forming a ring of coral separated from the island by a stretch of water. If the island disappears beneath the surface of the sea, a deep ring of coral, called an atoll, is left behind. The lagoon in the middle of the atoll is shallow, but the seaward side of the atoll is surrounded by deep water.

FRINGING REEF   **BARRIER REEF**   **ATOLL**   **LAGOON**

A volcano grows from the ocean bed. Corals grow around its slopes.

The volcano collapses and begins to sink.

The volcano vanishes, leaving behind an atoll.

HELPFUL SHRIMP

Some reef shrimps set up "cleaning stations" in conspicuous places where they are visited by a succession of fish. The shrimps seem to nibble at the mucus covering the fish's scales, but may also remove any external parasites. In return for cleaning up fish, the cleaner shrimps get a free meal. Cleaner shrimps often roll or sway their bodies as they move, which – together with their bright colors – may help to inhibit predators.

COLORFUL SLUGS

The brilliant colors found on many sea slugs warn predators that they have poisonous glands in their skins and will taste terrible. Some sea slugs even steal the stinging cells of corals and sea anemones and use them for their own defense. Another useful trick is to steal some tiny plants from the corals and farm them in their own tissues as an extra food supply.

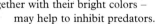

CORAL COASTS

Coral reefs contain the greatest variety of life of any marine community. The Great Barrier Reef of Australia contains over 3,000 different species of animal, and coral reefs generally support one-third of all the world's fish species. The superabundance of reef life is due to the year-round warmth and light as well as the plentiful supply of oxygen brought by the waves breaking over the reef. Coral reefs are ancient environments, having grown on Earth for some 450 million years. They have developed diverse lifeforms and complex interrelationships. A reef consists of a thin living skin of coral animals on top of layer upon layer of empty coral skeletons. A host of creatures, from fish, clams, and sea lilies to sponges, moray eels, and octopuses, feed or make their homes in the stony thickets and branches of the reef. The splendid colors of reef fish, may help species to recognize each other in the crowd!

BLENDING IN

The colors and patterns of the hawkfish blend in well with a background of coral. This predator often tucks itself into a fork of coral for greater camouflage. Then it can dart out to make surprise attacks on the fish it eats. This deep-water hawkfish is generally found on outer reef slopes at depths of 131 ft (40 meters) or more.

REEF BUILDERS

The Great Barrier Reef is the biggest coral reef in the world. It stretches for 1,430 miles (2,300 km) along the northeastern coast of Australia. At low tide, the reef seems to stretch to the horizon. During very low spring tides, the corals are exposed to the air for some hours. Corals are strange puzzling animals. They have a simple body plan with one opening for the passage of materials in and out of the body. Most corals live in colonies built up by one founder individual dividing to replicate itself over and over again.

YOUNG SEAWEEDS

EDIBLE PERIWINKLE

EDIBLE CRAB

LESSER OCTOPUS

Octopuses are intelligent predators that usually feed at night. They glide smoothly over the sea bed until they are within about 8 inches (20 cm) of their prey. Then they pounce, using jet propulsion to shoot forward and engulf the prey with their sucker-tipped arms. Octopuses have a highly developed sense of touch and can feel into cracks in between rocks to detect snails and shellfish hiding there.

PREDATORS & PREY

SHORE STINGERS

Anemones cannot move quickly to chase prey but their tentacles wave about in the water to grab tiny floating creatures or larger animals such as fish and prawns. Barbed stinging cells in the tentacles help to subdue the prey and paralyze it. The tentacles pass the prey down to the mouth opening and then spread out again to catch more food.

From razor-sharp teeth, bills and pincers, to barbed harpoons, stinging tentacles, and super-strong suckers, coastline predators are well equipped to catch their prey. Sea otters even use stones as anvils on which they smash open the tough shells of abalones or brush off the sharp spines of sea urchins. Hunting strategies include lying in wait for prey, like an anemone; ambushing prey, like an octopus; or chasing it like a seal or a cormorant. Shellfish are well protected by their strong shells but even these can be pried apart by a starfish or drilled through by a dog whelk or an oyster drill. Killer whales are true hunters. They hunt in groups and are able to overcome large prey, such as seals. The great baleen whales, such as blue whales and sperm whales, that filter plankton from the water are also classified as predators because they take living prey from their environment.

REEF SHARKS

These Caribbean reef sharks are hunting at the edge of a coral reef. Other reef sharks include the white-tipped and black-tipped reef sharks, tiger sharks, and hammerheads. A shark's sharp teeth help it to grip and cut up its prey. When the front teeth wear out, they are replaced by new teeth growing behind them. One shark can go through thousands of teeth in a lifetime.

BOXING SHRIMP

Mantis shrimps hunt for smaller shrimps, crabs, and fish. One pair of limbs has evolved to form "fists" rather like boxing gloves. These suddenly shoot out to deliver an impact equivalent to that of a 0.22 caliber bullet, to smash or stun their prey.

SPOTTED HUNTER

The leopard seal patrols coastal waters near a penguin colony, waiting for the penguins to dive into the water. It is a speedy seal with a long, flexible neck and a wide mouth for grasping penguins, seal pups, and other prey. Leopard seals pursue their prey under the water and then beat them against the surface to loosen the skin. This may peel right away from the body and slip up around the neck. Small penguins are usually swallowed whole.

SNAIL DRILLS

Rocky shore snails, such as oyster drills, dog whelks, and necklace shells, have to work quite hard for their meals. They pour a softening fluid over the shells of prey, such as mussels, barnacles, and limpets, and then drill a hole in the shell using their rasping tongue. A dog whelk may take two days to drill through a mature limpet or mussel shell, eventually producing a very neat, circular hole. Some of these drilling snails have a special tooth on the rim of their shells to help them pry open the double shells of bivalves.

SEASHORE FOOD CHAIN

PLANT & ANIMAL PLANKTON

↓

PEACOCK FANWORM

↓

FLOUNDER

↓

HERON

Many seashore animals are filter feeders, such as fan worms. They take in large quantities of water and pump it out through some kind of filter or sieve, which traps food particles for the animal to eat. Since they often filter both plant and animal plankton from the water, filter-feeders are called omnivores.

DEFENSE

Coastlines are busy, bustling places, full of creatures trying to eat each other. Some creatures, such as fish, octopuses, and squid can swim away from danger. Even scallops can clamp their shells together to escape the grip of a starfish. Hiding in burrows or among rocks or coral gives more sedentary animals a better chance of survival. Many are well camouflaged. The decorator crab makes its own camouflage from bits of seaweed and small animals such as sponges. The weedy sea dragon is a seahorse that looks just like a piece of seaweed. Thick, armor-plated shells protect shellfish against the weather and predators, while hermit crabs craftily move into an empty shell. More dangerous methods of defense include the sharp spines of sea urchins, the nasty nipping pincers of crabs, and the poisons of sea slugs. Animals that live in groups, such as fish in shoals or colonies of nesting seabirds, can warn each other of danger and help defend each other.

ROCKY BURROW

Piddocks use the fine teeth along the edge of their shells to drill through rock, turning one way and then the other, just like a drill bit. The piddock relies on its burrow to provide protection and is only partly enclosed within its shell. Piddocks cannot detect each other's presence in the rock and one piddock may drill slowly but surely straight through another.

COAT-OF-MAIL SHELLS

Chitons are sometimes called "coat-of-mail" shells because their shells are made up of eight overlapping plates joined and surrounded by a stretchy muscular band called a girdle. The many plates allow a chiton to bend its body easily and cling tightly to any rock surface, however uneven. This makes it difficult for predators to reach its soft body inside the shell. If a chiton is detached from a surface, it coils up to protect its soft body.

BORROWED HOME

Unlike their relatives the lobsters and true crabs, hermit crabs have a soft abdomen that is not covered by a hard, chalky shell. The crabs have to search out the empty shells of whelks and other sea snails and live inside them for safety. A hook on the end of the abdomen helps the crab keep a firm grip on the shell while it stretches its front legs to pull itself along. When the crab grows too big it moves to a bigger shell.

CRAFTY CAMOUFLAGE

Flatfishes are experts in camouflage. Their topsides match the color of the sea bed perfectly so they are almost impossible to see when they keep still. Many flatfish can change their coloring within minutes to match the sea bed. Their underside has no need of special coloring, so it is white or pale in many species.

PRICKLY MOUTHFUL

When puffer fish or porcupine fish are threatened, they inflate their bodies with water or air. This makes them too large for most predators to swallow. They also look more frightening. Porcupine fish and some puffer fish are also armed with sharp spines that stick out when they inflate, making an impossibly prickly mouthful.

SPINY PROTECTION

A sea urchin's long spines have three functions: protection, movement, and sometimes as digging tools to burrow into rocks. The spines are fixed to the shell, or test, of the urchin by muscles around raised areas which form ball-and-socket joints that allow them to move in all directions. The spines soon break off when the urchin dies and empty tests are all that is washed up on the beach.

NESTS, EGGS, & YOUNG

Many animals that live in the oceans, such as penguins, seals, and turtles journey to coastlines in the breeding season because they have to lay their eggs or give birth to their young on land. Warm, shallow coastal waters rich in food are ideal places for young fish to grow up; while nesting on coasts gives seabirds easy access to their food supply. Seabirds and seals feed and take care of their young, but many shore creatures leave their young to fend for themselves. They lay a lot of eggs because many will not survive to become adults. Female hermit crabs and some shrimps and crabs lay less eggs but carry them around. The common octopus even guards her eggs in her underground lair for about six weeks. The larvae of many invertebrate animals, such as hermit crabs, barnacles, and peacock worms may drift off with the plankton before settling down. As they grow, invertebrates with hard external skeletons, such as crabs, have to molt several times.

PREGNANT FATHER

This male seahorse is giving birth to babies that have developed in a pouch on the front of his body. The female lays her eggs in the male's pouch and the young emerge after two to seven weeks. As soon as they are born, the young seahorses must fend for themselves.

SEAL PUPS

The common, or harbor, seal gives birth to a single pup in late June or July, usually on sandbars, ledges, offshore islands, or ice floes and sometimes in the water. The pup is born well developed and relatively independent. It has an adult type of coat, unlike gray seals which have a white coat when they are born. The pups grow fast as they feed on their mother's rich milk.

MERMAID'S PURSES

Baby dogfish, which are a small kind of shark, develop inside an egg case called a mermaid's purse. At first the egg case is soft, but it soon hardens in the seawater. Tendrils anchor the egg case to seaweed to stop it from being swept away by the waves and currents. A large yolk sac inside the egg case nourishes the embryo for up to ten months before it hatches out to start a life of its own.

CORAL SNOWSTORM

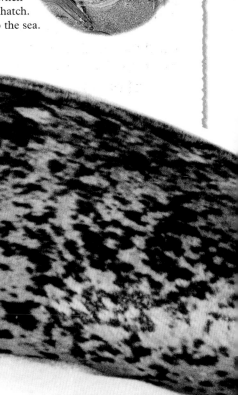

Corals release masses of eggs and sperm into the water, usually just after a full moon in spring or early summer. Some people have described the coral spawning as "an upside-down snowstorm." If a sperm fertilizes an egg from the same coral species, a tiny swimming larva develops. This drifts with the plankton before settling down to develop into an adult. Some coral eggs are fertilized while still on the adult, then brooded and released over long periods. Corals can also reproduce by dividing into two over and over again to produce exact copies of themselves.

BREEDING MIGRATION

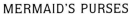

Small fish called grunions migrate in breeding shoals of many thousands to the sandy shores of Californian coasts during the high spring tides of March to June. The females wriggle through the wet sand and lay their eggs, which are then fertilized by the male. Two weeks later, when the tides are again very high, the young fish hatch. They squirm through the sand and into the sea.

PEOPLE & COASTLINES

DIVING

The underwater environment is alien to us and can only be truly experienced through diving. Aqualungs, developed in the early 1940s by Jacques Cousteau and Emile Gagnan, have revolutionized our exploration of coastal waters, both for recreation and scientific research. This diver is freeing a spiny lobster caught among the fronds of a giant seaweed called kelp. Coral reefs can be damaged by heavy boat anchors and by divers hunting for souvenirs.

For thousands of years, people have lived by the sea so they can fish and earn a living. Today, 60 percent of the world's population lives on, or near, the coast. Many of the world's major cities, such as New York, San Francisco, Sydney, and Shanghai are located on coasts where there are natural harbors or where rivers meet the sea. This is useful for trade and transporting goods carried by sea inland to other towns and cities. Mining operations can also be carried out in shallow coastal waters, such as dredging for tin in Thailand or mining diamond deposits off the Namibian coast. Drilling for oil and gas may occur off the coast. In some countries, farmland has been reclaimed from the sea or from coastal marshlands and shellfish or pearl oysters are farmed near the shore. Coastlines the world over are important places for tourism and leisure activities.

TOURIST JAM

On this crowded beach in Malia, Crete, there is hardly any space for the people, let alone wildlife. Birds and turtles that nest on beaches are often disturbed by the noise and bright lights in tourist areas. Beaches that are especially important for wildlife can be protected and new tourist developments can be planned to avoid areas of special value to wildlife.

FISHING COMMUNITIES

Colorful houses and fishing boats are a familiar sight on the coasts of the Caribbean islands, where fishing is a way of life for many. Around the world, small fishing communities are important to a coastal region as they supply food and act as a center of trade.

BEACH POLLUTION

Shorelines are narrow strips of land that can often become covered with rubbish left by people or washed up by the tides. Animals can injure themselves or become ill if they eat it. This fragile environment is full of life but can easily be damaged by the things we throw away.

FOOD FROM THE SEA

Many people who live near the sea work in the fishing industry or harvest other seafood, such as shellfish or octopuses. To make sure of a regular supply of oysters, this oyster farmer grows them in special cages in shallow inshore waters. The oysters grow best in clean, unpolluted water.

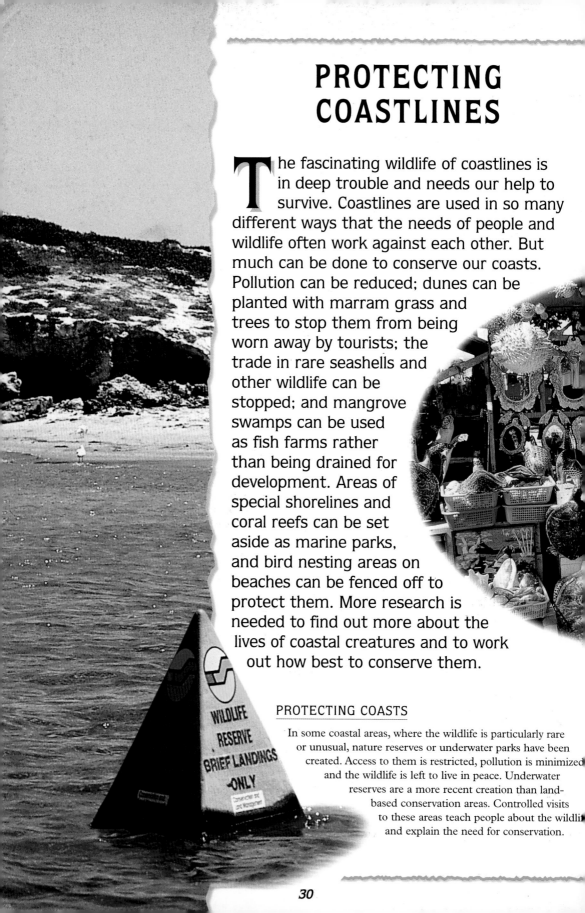

PROTECTING
COASTLINES

The fascinating wildlife of coastlines is in deep trouble and needs our help to survive. Coastlines are used in so many different ways that the needs of people and wildlife often work against each other. But much can be done to conserve our coasts. Pollution can be reduced; dunes can be planted with marram grass and trees to stop them from being worn away by tourists; the trade in rare seashells and other wildlife can be stopped; and mangrove swamps can be used as fish farms rather than being drained for development. Areas of special shorelines and coral reefs can be set aside as marine parks, and bird nesting areas on beaches can be fenced off to protect them. More research is needed to find out more about the lives of coastal creatures and to work out how best to conserve them.

WILDLIFE
RESERVE
BRIEF LANDINGS
ONLY

PROTECTING COASTS

In some coastal areas, where the wildlife is particularly rare or unusual, nature reserves or underwater parks have been created. Access to them is restricted, pollution is minimized and the wildlife is left to live in peace. Underwater reserves are a more recent creation than land-based conservation areas. Controlled visits to these areas teach people about the wildlife and explain the need for conservation.

RESEARCH

This manx shearwater has been
caught at night so it can be ringed
and its movements traced. These little
birds come to land only to breed and spend
the winter feeding at sea. Ringing these
shearwaters has shown that they migrate from
Britain to South America every year, flying at
least 460 miles (740 km) a day. It takes them
16 days or so to make this extraordinary journey.
The more we know about coastal wildlife, the
better we will be able to take care of it.

TRADE WARS

This souvenir stall in Java is selling the shells of endangered
turtles. If all tourists refused to buy such souvenirs less
rare animals would be taken from the wild.
It is not always easy to recognize
the rare species so it is best
to buy souvenirs made
by people.

COASTAL DEFENSES

Coastlines are
always changing
shape as the waves
and weather erode
the rocks. Some coasts
are being worn away while
others are being built up as
loose material is deposited to form
new and larger beaches. People often try to protect
coasts by building concrete or rocky defenses. But the sea
is a powerful force and in the long term there is little we can do.
In the future, many coastlines will be affected by global problems,
such as rising sea levels caused by global warming.

STRANDED WHALE

Whales may die at sea and be washed up on
the shore but live whales sometimes swim
onto the beach and strand themselves. Why
they do this is still a mystery but it may have
something to do with illness, pollution, or
problems with the whales' navigation system.
People sometimes save stranded whales by
keeping them covered with water and helping
them swim back out to sea when the tide
comes in. Watching live whales off the coast is
a growing tourist industry in many countries.

FIND OUT MORE

Useful Addresses

To find out more about life on the coast or conservation of coastlines, here are some organizations that might be able to help.

CENTER FOR MARINE CONSERVATION
1725 DeSales Street, Suite 600
Washington, DC 20036
(202) 429-5609 http://www.cmc-ocean.org/

CENTER FOR OCEANIC RESEARCH & EDUCATION
245 Western Avenue, Box 8
Essex, MA 01929
http://www.coreresearch.org/

FRIENDS OF THE EARTH
1025 Vermont Avenue, NW, 3rd Floor
Washington, DC 20005
(202) 783-7400 http://www.foe.org/

GREENPEACE
1436 U Street, NW
Washington, DC 20009
1-800-326-0959 http://www.greenpeaceusa.org/

OCEANIC RESOURCE FOUNDATION
P.O. Box 280216
San Francisco, CA 94128
1-888-835-9478 participate@orf.org

REEF ENVIRONMENTAL EDUCATION FOUNDATION
P.O. Box 246, Key Largo, FL 33037
http://www.reef.org/

SEACOAST INFORMATION SERVICES, INC.
135 Auburn Drive
Charlestown, RI 02813
(401) 364-9916 http://www.aquanet.com

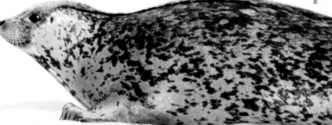

ACKNOWLEDGMENTS

We would like to thank: Helen Wire and Elizabeth Wiggans for their assistance. Artwork by Peter Bull Art Studio.
First edition for the United States, its territories and dependencies, Canada, and the Philippine Republic, published 2000 by Barron's Educational Series, Inc.   Original edition copyright © 2000 by *ticktock* Publishing, Ltd.
U.S. edition copyright © 2000 by Barron's Educational Series, Inc.
All rights reserved. No part of this book may be reproduced in any form, by photostat, microfilm, xerography, or any other means, or incorporated into any information retrieval system, electronic or mechanical, without the written permission of the copyright owner.
All inquiries should be addressed to: Barron's Educational Series, Inc.
250 Wireless Boulevard, Hauppauge, New York 11788   **http://www.barronseduc.com**
Library of Congress Catalog Card No. 99-68832   International Standard Book No. 0-7641-1076-4
Printed in Hong Kong.   Picture research by Image Select.
9 8 7 6 5 4 3 2 1

Picture Credits: t=top, b=bottom, c=center, l=left, r=right, OFC=outside front cover, OBC=outside back cover, IFC=inside front cover

Biofotos; 20/21b, 24/25b, 25br, 26/27t, 30/31c. Heather Angel; 2/3b, 3cr, 3tr, 4tl, 4bl, 4br, 5cr, 8br, 8bl, 9tl, 10/11(main), 12br, 12/13t, 13br, 14/15c, 15tr, 16bl, 16/17t, 18l, 18t, 19tl, 19tr, 23bl, 24tl, 31cr. Oxford Scientific Films; 9tr, 11bl, 11tr, 13c, 13tr, 14/15b, 16tl, 17ct, 19c, 23cr, 24bl, 26tl, 27tr, 27cr, 30l, 31t. Tony Stone; OFC, OFC(inset), IFC, 2/3t & OBC, 2l, 2ct, 4/5(main), 6l, 6/7b, 6/7t, 7tr, 7br, 10tl, 11cr, 12bl & OBC, 14l, 15bl, 14t, 16/17mp, 18/19b, 20tl, 21r, 21ct, 20cr, 22/23t, 22tl, 22bl, 25cr, 24/25t, 26/27b & 32, 28l, 28/29c, 28ct, 29tr, 29cr, 31br.

Every effort has been made to trace the copyright holders and we apologize in advance for any unintentional omissions.
We would be pleased to insert the appropriate acknowledgment in any subsequent edition of this publication.

BARRON'S